The Hazards Associate with Agricultural Silo Fires

Authored by: Alan Clark
John Kimball
Hollis Stambaugh

This is Report 096 of the Major Fires Investigation Project conducted by Varley-Campbell and Associates, Inc./TriData Corporation under contract EME-94-C-4423 to the United States Fire Administration, Federal Emergency Management Agency.

 FEMA

Department of Homeland Security
United States Fire Administration
National Fire Data Center

U.S. Fire Administration Fire Investigations Program

The U.S. Fire Administration develops reports on selected major fires throughout the country. The fires usually involve multiple deaths or a large loss of property. But the primary criterion for deciding to do a report is whether it will result in significant "lessons learned." In some cases these lessons bring to light new knowledge about fire--the effect of building construction or contents, human behavior in fire, etc. In other cases, the lessons are not new but are serious enough to highlight once again, with yet another fire tragedy report. In some cases, special reports are developed to discuss events, drills, or new technologies which are of interest to the fire service.

The reports are sent to fire magazines and are distributed at National and Regional fire meetings. The International Association of Fire Chiefs assists the USFA in disseminating the findings throughout the fire service. On a continuing basis the reports are available on request from the USFA; announcements of their availability are published widely in fire journals and newsletters.

This body of work provides detailed information on the nature of the fire problem for policymakers who must decide on allocations of resources between fire and other pressing problems, and within the fire service to improve codes and code enforcement, training, public fire education, building technology, and other related areas.

The Fire Administration, which has no regulatory authority, sends an experienced fire investigator into a community after a major incident only after having conferred with the local fire authorities to insure that the assistance and presence of the USFA would be supportive and would in no way interfere with any review of the incident they are themselves conducting. The intent is not to arrive during the event or even immediately after, but rather after the dust settles, so that a complete and objective review of all the important aspects of the incident can be made. Local authorities review the USFA's report while it is in draft. The USFA investigator or team is available to local authorities should they wish to request technical assistance for their own investigation.

For additional copies of this report write to the U.S. Fire Administration, 16825 South Seton Avenue, Emmitsburg, Maryland 21727. The report is available on the Administration's Web site at http:// www.usfa.dhs.gov/

U.S. Fire Administration

Mission Statement

As an entity of the Department of Homeland Security, the mission of the USFA is to reduce life and economic losses due to fire and related emergencies, through leadership, advocacy, coordination, and support. We serve the Nation independently, in coordination with other Federal agencies, and in partnership with fire protection and emergency service communities. With a commitment to excellence, we provide public education, training, technology, and data initiatives.

ACKNOWLEDGMENTS

The U.S. Fire Administration greatly appreciates the cooperation received from the following people and organizations during the preparation of this report:

Timothy G. Prather   Agricultural Safety Specialist
University of Tennessee Agricultural and Biosystems
Engineering, Knoxville, TN 37901

Bruce Warren   Director of Engineering
A.O. Smith Harvestore Products, Inc
DeKalb, IL 60115

Tom Aliek   A.O. Smith Harvestore Products, Inc.
DeKalb, IL 60115

Mr. Ken Musser   International Silo Association
Lafayette, IN

TABLE OF CONTENTS

Special Report:
The Hazards Associated With
Agricultural Silo Fires
April 1998

INTRODUCTION

Firefighting is by its very nature an occupation of considerable risk and potential danger. As a part of the United Stated Fire Administration's Major Fire Investigation Project, this Special Technical Report has been produced to help inform firefighters of the particular danger of fires in agricultural silos and the hazards that may be encountered in fire operations in and around these structures. Silo fires and explosions have been responsible for the deaths and injuries of firefighters and civilians. Since 1985, at least six firefighters have been killed in silo emergencies; three were killed in a silo fire explosion in Ohio, two in a different silo explosion in Georgia, and another was asphyxiated in a grain storage facility in Iowa. The lethal potential is also underscored by the fact that many civilian workers have lost their lives in non-fire situations in grain or silage storage facilities.

Some silo emergencies can be immediately life threatening and all can be dangerous to firefighters if appropriate safety precautions are not followed. This special report is meant to provide information to prepare for and safely extinguish fires in agricultural silos.

The most common purpose of a silo is to store and preserve foodstuff for livestock. Whether the product is ensiled corn, hay, or bermudagrass, adequate quantities of high quality feedstock is the basis of a profitable milk and livestock production. Silos allow the controlled and flexible harvesting, storage, and disbursement of high quality feedstock. Hence, silos are an essential component of farm operations.

The use of silage has actually grown in popularity in the past 50 years because it maximizes the yield of nutrients from available land, decreases the cost of feed, lowers harvest losses, and increases the quality of the foodstuff. The use of silage can make the difference between a profit and a loss. Even with the capital outlays required for equipment, the mechanization of harvesting and storage reduces labor costs and related requirements. A trend toward greater silage use is likely to continue as more dairy and livestock producers graduate to year-round feeding of storage feedstocks and as livestock operations increase in size.

As with all emergency operations, no single guide can cover every possible scenario. Results vary with each situation. Experience and proven methods should be used in addition to guides in extinguishing silo fires. However, each silo fire needs to be analyzed and decisions made based on the specific factors of the individual fire. The information in this special report can assist fire officers in the strategic and tactical planning for agricultural silo fires. As in any fire size-up and decision-making process, the most important factor is personnel safety.

1

SCOPE OF THE PROBLEM

In today's farm economy, the silo is an integral component of a feed management system. Silo fires, therefore, can be very costly to farmers. In large modern operations, the amount of silage required is carefully calculated based on the farmer's cost per pound of meat or per gallon of milk produced. The interrelationship of yield of silage per acre, nutrient value per ton, required output of meat or milk, and cost per ton of silage could be upset by a catastrophic fire that destroys significant amounts of silage. Another concern for the farmer is the difficulty of obtaining silage from alternate sources if silage is destroyed. The cost is considerably more per ton since silage does not lend itself well for transfer and shipment over long distances or extended times. The loss of silo operations can portend serious financial trouble.

In some respects, silos can be considered the rural version of a highrise building. Similar challenges include the elevated area in relation to the rest of the farm buildings, limited access, and labor-intensive operations requiring coordination of resources. Like highrise fires, however, silo fires can be managed with an organized and methodical approach, utilizing on-scene resources and the information gained in pre-incident planning. Silo fires can be life-threatening in addition to being resource intensive. The following incidents indicate the potential dangers of silo fires.

Marshallville, OH, August 27, 1985

Three firefighters were killed while they were attempting to extinguish a fire in an oxygen-limiting silo by spraying water into the silo. They were on the top of the silo directing a water stream when an explosion lifted the concrete roof of the silo approximately four feet in the air, throwing the fire-fighters to the ground and into the silo.

By directing water into the top of the silo from the roof, air was entrained from the hose stream into the area of incomplete combustion. This provided oxygen into an atmosphere of high heat and sufficient fuel, which then causes an explosive backdraft.

Bostwick, Morgan County, GA, August 5, 1993

Two firefighters were killed when they applied water and foam to a fire in an oxygen-limiting silo. The explosion blew the roof off, sending one firefighter to the ground over 100 yards away and the other through the roof of a nearby metal building. Two firefighters on the ground were injured by debris. The top 15 feet of the silo were severely damaged by the explosion and an adjacent silo was dented by the debris, attesting to the force of the blast.

SILAGE PROCESS AND FIRE RISK

In the process of storing silage a plant material, "forage," is harvested and stored in a container, "silo," where it is preserved for livestock use. Silage is a preserved forage that may consists of grain crops such as chopped corn or sorghum; or legumes such as alfalfa, bermudagrass, or other hay crops. A fermentation process develops in the silo that preserves the silage.

The forage is chopped into small pieces and stored at 45 to 70 percent moisture and packed tightly in the silos. The material must be packed in order to force air out of the mass of forage for the preserving process to work properly. As the still-living forage is stored, it continues to respire, consuming the oxygen in the ambient air. As the air is used up, anaerobic (without air) bacteria multiply and consume plant sugars to produce lactic acid which raises the acidity of the silage. This action kills the bacteria and halts the fermentation process. At this stage the silage is stable.

If oxygen becomes available from breaches in the silo structure, the aerobic bacteria growth begins again. This growth consumes oxygen, uses nutrients, and causes heating. The consequences of this range from loss of nutrients and less efficient feedstock to spontaneous heating sufficient to cause ignition. If the silage is too dry when stored, is allowed to dry out, or is not densely packed, and if the oxygen is available due to poor structural conditions, spontaneous ignition is possible.

TYPES OF SILOS

In order to conduct fire planning for silos, fire departments must recognize the basic silo types as well as the construction features, identifying characteristics, and hazards associated with each type.

Silos are categorized generally as horizontal (bunker) or vertical (tower).

Horizontal Silos

Horizontal silos can be either an above-ground bunker, a trench, or a large stand-alone mound of silage contained in plastic. These trenches or above-ground bunker silos are generally used for large herds requiring 400 tons or more of silage. They are constructed by mounding the soil around a trench or by sinking concrete walls into a trench. These "structures," present the hazards of below-grade and confined space operations and the production of silo gases. Spontaneous ignition also can occur.

Vertical Silos

Vertical or tower silos are upright structures ranging in height from 30 to 120 feet with the 60- to 80-foot range being the most common. Silos may measure from 12 to 30 feet in diameter, though most are between 20 and 24 feet.

The two most important types of vertical silos are conventional and oxygen-limiting or "controlled atmosphere." The differences are basic, but the ability to distinguish between the two can literally make the difference between life and death in the event of a fire. Diagram 1 illustrates the basic differences between the two types of silos.

Conventional--Conventional silos are by far the most common type of silo. They are massive structures. An average size (20' x 60') concrete stave silo can weigh 70 tons and hold seven times that weight in silage, depending on the product stored. These silos are built on a reinforced concrete pad or base. It is not unusual for several silos to be erected side-by-side servicing a command feed room.

The most common conventional silo is one constructed of concrete staves. Staves are curved rectangular-shaped concrete blocks each weighing 70 pounds or more which are held in place by reinforcing bands made of high strength steel.

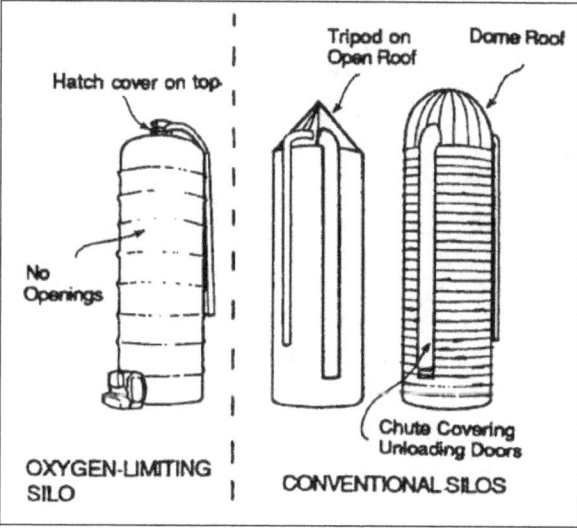

Diagram 1. Types of Silos

The bands are threaded on each end and bolted together to form a circle around the exterior of the staves. Conventional silos may also be constructed of wood, poured concrete, tile blocks, or steel.

Main Features--There are two main characteristics of conventional silos: an unloading chute that runs the vertical length of the exterior of the silo, and an open top or loosely constructed dome. The chute is about 3' in diameter and is used to unload and drop silage into a conveyor belt or feed room. The unloading chute can be steel, sheet metal, or fiberglass. Unloading doors are located every few feet in a column configuration similar to a floor opening. The unloading doors, located inside the unloading chute, are most commonly wood with steel hinges, handles, and ladder rungs bolted through the door.

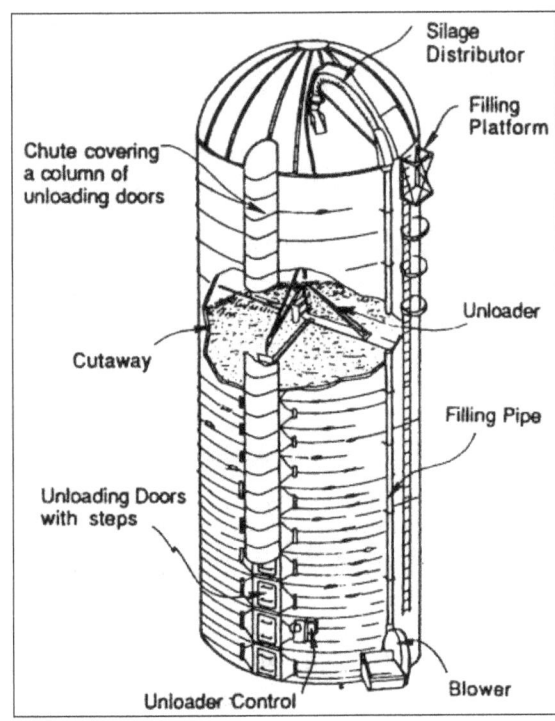

Diagram 2. Features of Conventional Vertical Silo

The dome or roof cap can be constructed of fiberglass, aluminized steel, or other similar materials. Those silos with a roof cap or dome generally have a fill tube and loading door on the side of the top of the dome. Many concrete stave silos have a metal fill platform at the top of the outside ladder. Many of the silos also have a steel safety cage surrounding the exterior ladder. Ladders and loading door platforms are all less frequently found on the block and wood silos. Conveyor belts many times are located directly below the unloading chute to transfer silage to feed bunks in nearby feedlots. (See Diagram 2.)

Oxygen-Limiting--One type of vertical silo is an "oxygen-limiting" or "controlled atmosphere" silo. The purpose and function of this type of silo is just as the name indicates: to limit oxygen as the silage is stored. While the conventional silo is constructed to minimize the free flow of air into the storage area, the oxygen-limiting silo is designed as solid construction to be nearly air-tight. They are commonly constructed of steel shells with an inner layer of glass bonded to the steel to protect it and augment the insulation properties. The outer covering may be a finish of a ceramic and enamel coating. Poured concrete and fiberglass silos also are in operation today. Conventional silos of poured concrete construction can be converted to function as oxygen-limiting silos. The danger here is that these silos appear to be conventional, but they are actually oxygen-limiting and must be treated as such. The primary characteristics will be no silo openings, no unloading chute, and a fill pipe extending into the top of the dome. Conversion will also be identified by the bottom-unloading configuration.

Main features--The primary identifying feature of oxygen-limiting silos is the absence of an unloading chute on the exterior of the silo. Common characteristics of oxygen-limiting silos can also include a fill door located in the center of the top of the silo roof, a roof hatch rather than a loading door, a center fill pipe attached to a blower or fill pipe, or a single unloading door at the bottom of the silo. Oxygen-limiting silos primarily use bottom unloading--filling from the top and removing silage from the bottom. Most oxygen-limiting silos also have a steel ladder and safety cage leading up to and across the roof to the center fill door. (See Diagrams 3 and 4.)

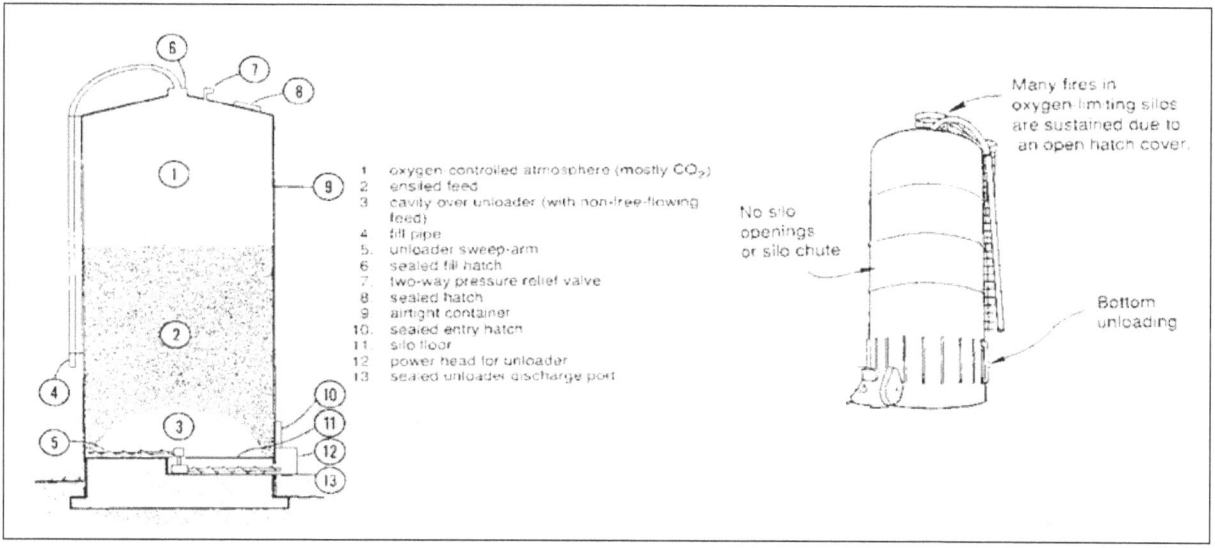

Diagram 3 and 4. Features of an oxygen-limiting silo

Other features that may be found on oxygen-limiting silos include two-way pressure relief valves on the roof, roof hatches, and access doors at ground level with safety interlocks.

Two-way pressure relief valves compensate for interior pressure changes caused by temperature variations. (Diagram 5.)

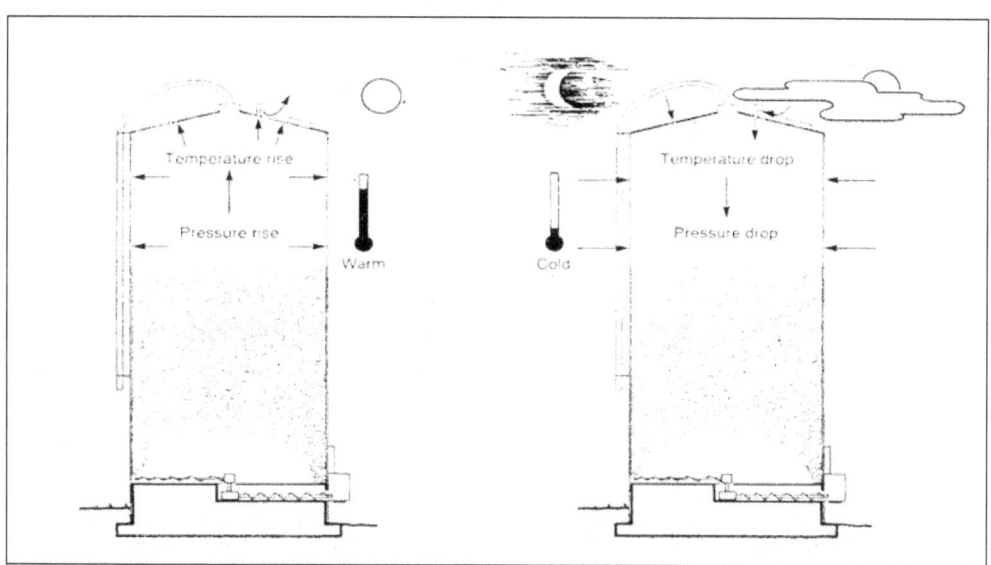

Diagram 5. Oxygen-limiting silos must "breath" to compensate for temperature and pressure changes

Oxygen-limiting silo may also be equipped with a "breather bag" arrangement designed to compensate for pressure changes in the interior of the silo. (See Diagram 6.)

Warnings or safety decals may be posted on oxygen-limiting silos. These are primarily intended for farm workers in a routine setting. Firefighters should consider fires or incidents in an oxygen-limiting silo as an imminently hazardous situation, which could result in death or serious injury.

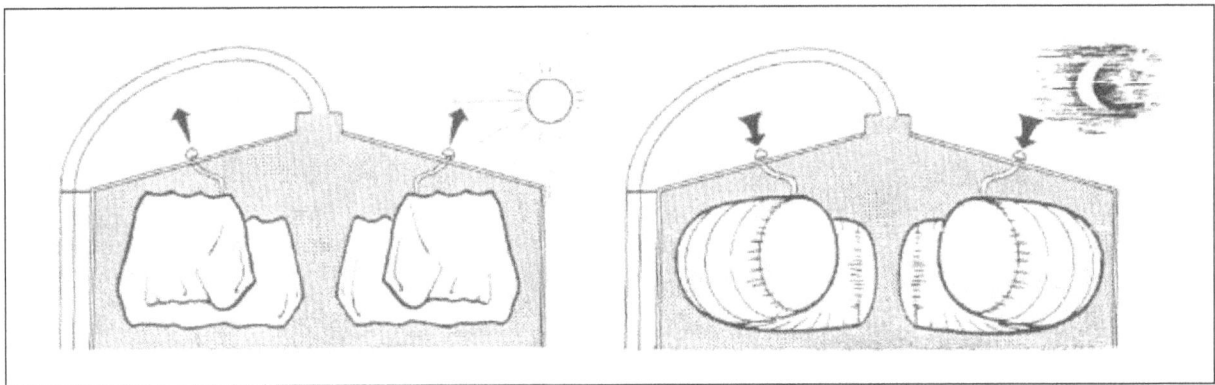

Diagram 6. "Breather Bag" system in oxygen-limited silo

Conversion silos--Conversion silos are those that were constructed originally as conventional type silos and have been modified to be oxygen-limiting. Modifications consist of reinforcing masonry blocks and joints to prevent any air exchange. Another conversion type is that which has been constructed using concrete or fiberglass, but that which is of such structural integrity as to control the atmosphere and, limit the oxygen in the silo. (Diagram 7.)

The same cautions and recommendations apply to conversion silos as to the oxygen-limiting silos: **Do not attempt to enter, breach, or otherwise attempt to extinguish fire in a converted conventional to oxygen-limiting silo.**

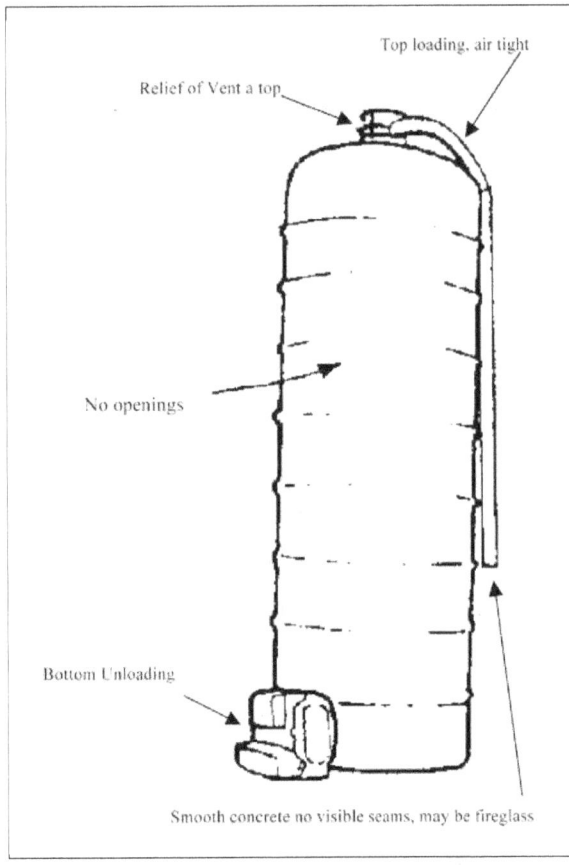

Top loading, air tight

Relief of Vent a top

No openings

Bottom Unloading

Smooth concrete no visible seams, may be fireglass

Diagram 7. Conversion type of silo

FIRE CAUSES

Different construction characteristics and contents of the two primary types of silos can result in different fire causes and personnel hazards. However, some of the universal causes are:

Spontaneous ignition--The most common cause of fire in silos is spontaneous ignition. This usually occurs when the silage is stored at improper moisture levels or the silage particles are not properly sized. Care must be taken to ensure the proper moisture and distribution of silage. Silage contains living cells that continue to respire (breath) using oxygen-consuming plant sugars and producing heat. Generally, as the oxygen is consumed and the natural fermentation process slows, the silage reaches a stable state. This occurs approximately three weeks after storage in the silo. However, if the anaerobic process continues for a long period, the temperature of the contents can rise to the temperature necessary to produce spontaneous ignition. Documented cases of spontaneous ignition have been recorded in silage that has been stored for over two years.

Fires are much less common in "oxygen-limiting" silos than in "conventional" silos, but the death and injury potential is much greater. That is because the former are designed to be airtight. If any of the latches or doors is left even partially open or the structural integrity is breached through leakage, air will enter the storage area and hasten the spontaneous combustion process. Toxic and explosive carbon monoxide is produced in much the same process and principle as a backdraft. If an ignition source is produced or the balance of the incomplete combustion is upset, an explosion is likely. These backdraft-like events have resulted in fire fighter fatalities as noted in the Incident Summary.

Lightning--As silos are generally the tallest structures on the farm, they are prone to severe damage from lighting strikes. Lightning rod protection systems can prevent damage to the structure, but damage and fire can still result if lightning strikes hit the facility's electrical system.

Mechanical heat of friction--Loading or unloading belts, pulleys, shafts, and other mechanical apparatus are subject to mechanical failure and overheating.

Electrical--New motors used in silos and feed rooms have safety devices; but many older electrical motors have been modified, and supporting wiring and circuitry may be substandard. Older or misapplied units such as these are prone to arcing and overheating.

CASE STUDIES

The following is a selection of sample incidents from 1991-1996 involving fires in silos. The case studies represent a range of silo-related fires and explosions. They illustrate the dangers and potential outcomes of silo fires, as follows:

1. A grain dust explosion in a converted oxygen-limiting silo.

2. A backdraft explosion in an oxygen-limiting silo.

3. A fire originating in a conventional silo that was successfully contained and extinguished.

4. A fire originating in a conventional silo that extended out of the silo, destroying several buildings.

5. A backdraft explosion in an oxygen-limiting silo in a nonagricultural setting.

Case Study 1. Explosion in Conventional Converted Oxygen-Limiting Silo, Louiston, Minnesota

The Explosion

On Tuesday, October 15, 1996, at approximately 3:15 p.m., a silo explosion rocked a farm near Louiston, Minnesota. The explosion occurred on a sunny day with no thunderstorm activity. The owners were picking corn and placing it in a 21-feet-diameter by 30-feet-tall fiberglass oxygen-limiting silo. The corn blown into the silo was high moisture (24 percent) and was brought directly from the field to the silo. The silo was approximately 18 years old.

A person hired to fill the silo was directing the corn through a metal fill pipe into the top of the fiberglass silo. The top hatch had not been opened prior to ensiling the corn. According to a second hired man, the explosion occurred about two minutes after the filling process began. Eyewitnesses said the explosion shook dust off the rafters in the feed room next to the silo and was heard by others working in the farmyard.

Scene Examination

Examination revealed a large crack in the upper third of the fiberglass silo. The break was located in the vapor space above the corn which was stored in the bottom portion of the silo.

In previous practice when the silo was being filled, the top hatch was left open to allow corn dust to escape the silo. This time the hatch was closed, promoting conditions favorable for a grain dust explosion:

- a supply of oxygen from air inside the silo;

- a fuel source (grain dust-corn dust) added in the filling process;

- dry grain dust suspended in the air; and

- proper ratio of dust to air located in a confined space.

The only thing missing was an ignition source.

Ignition Source

No electrical, heat-producing, or open flame devices were located or being operated at the top of the silo. No lightning storms were in the area at the time if the explosion. Based on the site investigation, the source of ignition was determined to have been a static spark generated by blowing corn through the undergrounded metal fill pipe of the silo.

Corn dust has the following characteristics according to section 5, chapter 9, of the 16th Edition of the Fire Protection Handbook published by NFPA.

- explosibility index 6.9 (Scale of 0-10+);

- explosion severity 3.0 (Scale of 0-2+); and

- maximum explosion pressure 113 PSIG.

The above characteristics of corn dust create the potential for a "severe" explosion. Eyewitnesses recalled hearing the explosion, then seeing the crack or break in the silo. Grain dust explosions can generate overpressures in excess of 100 PSI. For comparison, wood-framed walls can normally withstand pressures of 1-2 PSI, and concrete structures can withstand 2-8 PSI.

Conclusion

Grain dust explosions can occur within an oxygen-limiting silo. Firefighters must use caution if presenting conditions indicate the possibility of a dust explosion. Generally, oxygen-limiting silos prevent the conditions necessary for a dust explosion. However, if the balance is upset, a dust explosion can occur. **The force generated by such an explosion is sufficient to damage heavy structural members and cause injury or death in a substantial surrounding area.**

Damage to silo from explosioncrack in vapor space

Case Study 2. Explosion in Oxygen-Limiting Silo, Moville, Iowa

The Explosion

On Tuesday, April 29, 1994, at approximately 9:30 a.m., an explosion occurred at a farm near Moville, Iowa, in an oxygen-limiting silo being used to store corn silage and haylage. The owners were attempting to unload the silo contents from the bottom in order to reach a level where silage had been burning slowly for over a week.

The explosion occurred at approximately 9:30 a.m. and ripped the roof off of the silo. Although workers were in or near the feed room, no injuries were reported.

Scene Examination

An investigation into the explosion revealed that the roof of the silo was blown off and landed about 100 feet from the silo. Bolts were torn through the upper walls of the silo and the steel sheets were left crinkled in the upper portion of the structure. The breather bag was found to be broken. This breach in the bag was the source of air (oxygen) which allowed a spontaneous combustion fire to develop in the sealed storage unit.

The explosive gas in this incident was believed to be carbon monoxide (CO), which has an explosive range of 12.5 percent to 74 percent. The ignition temperature of carbon monoxide gas is approximately 1130 degrees Fahrenheit.

Upon ignition, the pressure wave proceeded up the silo with enough force to explosively expel the roof. The resulting vacuum inside the silo crushed the side of the silo due to pressure differential created by the explosion.

Ignition Source

The spontaneous combustion fire within the haylage and corn silage produced a high enough temperature to ignite a concentration of carbon monoxide gas within the vapor space of the oxygen-limiting silo.

Conclusion

Explosions can occur in oxygen-limiting silos. The force generated by such an explosion is sufficient to damage heavy structural members, and propel heavy objects great distances. In this case, a damaged breather bag allowed oxygen into the sealed unit and spontaneous combustion occurred. The fire caused by the spontaneous combustion ignited the carbon monoxide mixture in the silo, causing the backdraft-like explosion. The explosion generated sufficient force to propel heavy objects over one hundred yards.

Deformity of silo shell due to explosion

Travel distance of silo roof due to explosion

Case Study 3. Fire in Conventional Silo, Elgin, Minnesota

The Fire

On March 10, 1995, a fire believed to have been caused by spontaneous combustion was reported by a farmer to the Elgin Volunteer
Fire Department.

After determining that the fire was contained and not subject to
spreading, members of the fire department met to plan their attack.
After pre-planning session of one hour and a review of safety procedures, the Elgin Volunteer Fire Department then responded with
one pumper, one tanker, and two equipment support trucks.

Fireground Size-Up

Upon arrival at the scene, fire was observed in two areas from
the loading platform. One spot of open burning on the surface of
the silage was observed directly below the loading door. A second
location of open burning was observed in the area of the unloading doors and chute. In this area, fire had burned out a "U"- or
"V"-shaped pattern around the unloading doors and had burned
out several of the upper doors as well. Heavy ash residue and glow-
ing embers could be seen in the area of the unloading chute doors.

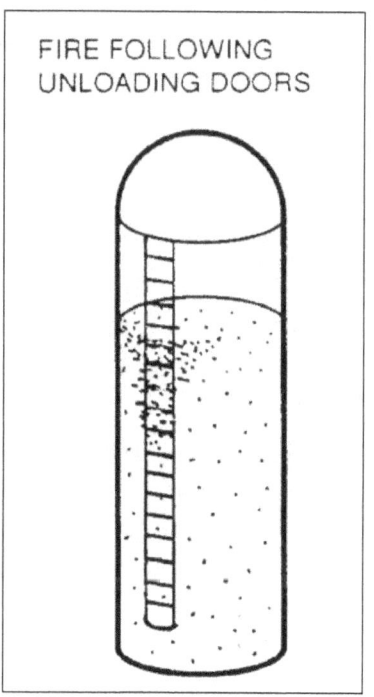

FIRE FOLLOWING
UNLOADING DOORS

**Diagram 8. Fire near
unloading doors**

Moderate to heavy smoke filled the vapor space and dome area. The conventional concrete stave silo was more than 50 percent full of silage.

The unloading chute was clear of smoke and a positive up draft was noted with the loading door open at the top of the silo.

Fire Extinguishment

An inch-and-one-half line was advanced up the exterior of the silo to the loading door platform. A firefighter used the line to attack the visible burning in the surface area. A second inch-and-one-half line was advanced up the unloading chute with a silo fire probe attached. This enabled firefighters to attack the fire from inside the unloading chute, saving them from actually having to enter the silo.

Extinguishment commenced in the silo interior with knock-down of burning contents achieved by using the probing nozzle. The contents were then shoveled through the chute, falling to the base of the silo. The burning silage was doused with water and remaining fire was wet down. During probing activities inside the chute, heavy smoke, heat, and steam were produced inside the silo. This was quickly dispersed by the use of positive pressure ventilation in the unloading chute. After one hour, enough fire had been extinguished to reduce the temperature within the vapor space of silo to approximately ambient air temperature.

Silo Entry

Due to a large cavity of burned-out silage around the unloading doors, a six-foot fire department ladder was placed from the unloading door over the silage to the silo unloader. This ladder was secured to the unloading door opening and to the unloader. Firefighters entering the silo were wearing safety harnesses and secured to anchor points above. At this time, the atmosphere inside the silo was clear with little evidence of smoke or steam.

Using a temperature probe, surveillance was conducted in the area of the unloading doors and the exterior walls of the silo. Crews worked in 30-minute shifts. They were monitored by a safety officer who communicated via portable radio from his vantage point on the loading door platform. No smoke or fire could be found at the conclusion of the fire ground operation. However, it is believed that the fire was only 95-98 percent extinguished because, once the unloading process began, spots of fire were discovered and wet down.

By 9:00 p.m., no further fire could be found and the owner was advised he could begin unloading the silage to a level one door below the lowest area of fire.

Conclusion

Fires in conventional silos can be safely extinguished. There is generally time to pre-plan fire ground operations in conventional silos. Positive pressure ventilation can be used with good effect. Firefighters should always anticipate hot spots and burning embers after the primary extinguishment has been completed.

Case Study 4. Silo Fire Extended to Barn, Delmar, Iowa

The Fire

On May 20, 1991, at approximately 4:30 p.m., a fire was discovered by farm owners who were feeding cattle from the east silo. Responding firefighters discovered the first evidence of fire in the east silo feed room. The fire quickly spread to the 28 x 120-foot cattle shed directly to the south of the silos. A 16 x 88-foot feed bunk south of the cattle shed was severely damaged and an adjacent 50 x 60-foot dairy barn located southeast of the cattle shed was totally destroyed by fire as well. The contents of one of two 18 x 88-foot concrete stave silos also was lost.

Ignition Source

Examination of the barn, feed bunk, cattle shed, and west silo failed to reveal any evidence of an ignition source for the fire. Examination of the east silo showed burn damage to the doors in the unloading chute to be more extensive on the inside of the silo than on the exterior surfaces of the wood doors. The fire's origin was indicated to have been inside the silo.

Evidence discovered during the investigation was consistent with the initial cause declared as spontaneous combustion of the silage in the east silo. Hot embers dropping into the feed room during the unloading of the burning silage are believed to have spread the fire. These hot embers ignited combustible materials in the feed room. Fire then spread from the feed room to the cattle shed dairy barn and east silo. Silage in the west silo was found undamaged.

Conclusion

Spontaneous ignition in conventional silos can produce a slow developing fire which can be contained inside the silo if undisturbed. If the fire extends outside the silo, the blaze can move rapidly and cause extensive exposure fires.

Fire originated in the silo on the right, then extended to feed room dairy barn, cattle shed, and feed

Destruction of farm buildings due to extension of spontaneous combustion-caused fire in silo

Case Study 5. Silo Fire and Explosion, Sawdust Storage, Statesville, North Carolina

The Fire

On December 21, 1997, at 8:35 a.m., fire units responded for a report of a fire in a silo to a facility that manufactured wooden reels for cord and cable.

The silo in question was an oxygen-limiting silo of the type designed for storage of agricultural silage. It had been converted for use as a storage and transfer container for wood waste and by-products of the manufacturing process. Three fire companies responded and proceeded to apply water to the deep-seated fire in the silo. A crew was placed on the dome of the silo and operated a 1-3/4" attack line with a 30-degree fog pattern directed into hatch openings. After approximately two hours, the decision was made to remove access plates on the bottom of the silo and manually remove the wood products and complete the extinguishment of the fire. As the last bolt holding the plate was removed, both the crew on the top and the firefighters at the bottom of the silo noticed an inrush of air and a dull thump. This was followed by a loud, low-order explosion and fireball emanating from the area near the access hatch.

The Explosion

The explosion blew the roof structure and three firefighters from the top of the silo. One firefighter went straight up and landed in the interior of the silo, one was blown onto the top of a nearby shed, and the third was entangled between the top of the silo and the ladder cage from where he had been operating. Firefighters on the ground were propelled by the explosion away from the hatch opening, narrowly escaping injury. Moments before, the fire chief had removed personnel from the immediate vicinity of the opening.

A two-hour rescue operation was required to extricate the firefighters from inside and on top of the silo. The other firefighter was retrieved quickly from the shed. The firefighter who was dropped into the silo was hospitalized for second-degree burns to the face and neck. The one who was trapped on the top was treated and released for burns and contusions. The firefighter who landed on the shed suffered serious knee and shoulder injuries requiring surgery.

Conclusion

Explosions can occur in oxygen-limiting silos regardless of the products stored. Personnel should not be sent to the roof of an oxygen-limiting silo. Fire departments carefully evaluate the risk factors and consider defensive operations and exposure protection.

Damage to silo hatch at base; one firefighter blown onto the roof of shed at right.

One portion of the roof traveled approximately 400 feet to the right.

HAZARDS ASSOCIATED WITH SILO FIRES

The most serious and unique hazard of a silo fire is the explosion potential present in oxygen-limiting silos, but firefighters face other significant hazards in both types of silo configurations as well.

1. Combustion Explosion

As previously noted, oxygen-limiting silos pose a very real danger to firefighters from an explosion standpoint. Is a fire occurs in an oxygen-limiting silo from spontaneous combustion, carbon monoxide is produced and can collect in the vapor space of the silo. As the flammable range of carbon monoxide is very wide, from 12.5 to 74 percent, the potential for a backdraft-like explosion is very high. If exterior attempts to cool down the silo involve directing water through roof openings, air is drawn in from the outside, which may bring the flammable range concentration of CO into the right proportion for an explosion. The source of ignition is the heat of the spontaneous combustion that is already occurring. Any attempted entry into oxygen-limiting silos may cause rapid combustion explosion.

2. Grain Dust Explosion

Though normally not considered a silo fire hazard, grain dust in oxygen-limiting silos have been documented as the cause of some explosions. Case study number 2 is an example of this. If the dust inside the silo becomes suspended as result of fire operations and is ignited by the heat of a smoldering fire, a dust explosion can occur.

3. Confined Space Dangers

All silo entries should be treated as confined space entry because they fit all criteria of the definition. Firefighters need to be aware of the following possibilities:

- Toxic gases, depleted oxygen levels, and flammable gas mixtures may be present.

- Electrical energy, potential mechanical energy (such as hydraulic pressure), or energizable machinery may endanger firefighters if such sources of energy are not controlled, locked out, and tagged out.

- Material may be "bridged." Bridging occurs when the fire burns, creating a void in the silage beneath what otherwise appears to be solid material. Bridging may also present as unevenly piled material, or appear as unstable ledges or overhangs of silage.

- Material may shift or be drawn to the bottom of the silo, causing anyone inside to be engulfed and trapped in the silage. A firefighter may suffocate from this type of entrapment, even if wearing self-contained breathing apparatus.

- Vision may be obscured.

- Communication may be difficult.

- Small clearance areas may hamper mobility, especially while wearing breathing apparatus.

4. Toxic Gases

A number of hazardous atmospheric conditions are possible in silo incidents. Toxic gases may be produced by the reaction of the silage, incomplete combustion, or oxygen depletion. Among the specific gases are:

Nitrogen dioxide--Exposure to nitrogen dioxide (NO2) can be fatal and is linked to a serious lung disease, commonly referred to as "silo filler's disease." (This is a term for chemical pneumonitis caused by exposure to nitrogen dioxide found in silage.) Nitrogen is produced when the oxygen is consumed in the fermentation process and the nitrates present in the plant material are released to form one of the oxides of nitrogen. As it escapes, it combines with the oxygen in the air to form the very toxic nitrogen dioxide. When NO2 is inhaled, it combines with the moist tissue of the respiratory system to form nitric acid. Strong concentrations of NO2 can create instantly lethal pulmonary edema due to the nitric acid in the lungs and throat. Reactions to moderate concentrations can cause pulmonary edema up to 30 hours later after the initial exposure, with the possibility of relapse in two to six weeks. **The most dangerous period for encountering harmful nitrogen dioxide gas concentrations is from five to eight days after the silo is filled.**

A particularly dangerous property of nitrogen dioxide is the synergistic affect with carbon dioxide which is also a byproduct of the silage process. At low level concentrations, nitrogen dioxide may cause little immediate pain or discomfort. The carbon dioxide present will cause an increase in respiration rates especially during a state of exertion (such as operating at a silo fire). This causes the individual exposed to inhale more rapidly and deeply, which brings the nitrogen dioxide deeper into the respiratory system without the warning of physical discomfort. If a firefighter inhales dangerous amounts of nitrogen dioxide, there is a high potential of developing fatal pulmonary edema, even well after the incident.

Nitrogen dioxide gas may be visible as a layered reddish/yellowish/brownish haze having a strong bleach-like odor. It is very toxic, having a Permissible Exposure Limit of five parts per million, and is heavier than air (vapor density is 1.58). Nitrogen dioxide gas may be present just above the top of the silage level and, being much heavier than air, may travel down the unloading chute and concentrate in low places such as the feed room, pits, or other below-grade areas. Other indicators of nitrogen dioxide gas are:

- strong burning sensation in the nose, chest, and throat;

- an acrid bleach-like odor;

- unnatural breathing of livestock;

- the presence of dead insects, animals or birds, especially in the lower levels of the silo/feed rooms; and

- the presence of a yellow stain on silage, wood, or other materials.

Any of these warnings may indicate the presence of NO2 and extreme caution should be exercised. The use of self-contained breathing apparatus is indicated.

Carbon dioxide--Carbon dioxide gas is produced in quantity in the silage fermentation process. It is odorless, colorless, and tasteless, and heavier than air. Carbon dioxide is non-toxic, but displaces the air and lowers the oxygen level potentially to the point that respiratory distress can occur. Strong concentrations of CO_2 can cause rapid asphyxiation due to oxygen insufficiency.

Carbon monoxide--This gas is a product of incomplete combustion and is a potential hazard in oxygen-limiting silos. It is very toxic and causes asphyxiation by preventing oxygen from being carried in the red blood cells. It has a Permissible Exposure Limit of 50 parts per million and is lighter than air. The primary hazard of this gas in silo environments is, however, the threat of explosions due to the high combustibility of carbon monoxide.

5. Electrical

Firefighters face numerous electrical hazards in silo environments. Electrical cables run up the unloading chute to an electric motor attached to the unloader. Conveyors, feed systems, and lighting in the feed room all pose potential electrical hazards. Consequently, electric power should be controlled prior to entry of fire personnel, the application of water, or other fire operations.

Another electrical hazard is the presence of overhead wires or transformer poles. In the farm environment, these may be poorly marked, low to the ground, and in poor condition. Every year, an average of 62 U.S. farm workers are electrocuted in accidents.

6. Climbing/Falling

Many silos extend up to 100 feet in height, thereby presenting a climbing hazard. Although most silos have ladders with a safety cage, the condition of the ladder and supports should be examined before they are used. Firefighters should check the condition of the ladder and cage (rungs, side rails, nuts, bolts, rivets, and other fasteners) for signs of rust, missing pieces, age, and other signs of weakness or failure. They should avoid using the ladder if it is not in good condition.

Falling is a potential inside the unloading chute because silage, mud, water, or other debris can make for very slippery conditions. Unloading doors should be checked for secure closures and each hand hold or footstep should tested to ascertain structural soundness.

The entire operating area may be subject to mud, animal waste, and other debris.

7. Chutes

The condition of the chute itself should be observed in the conventional concrete stave silo. The chute is normally connected to the silo by the use of nuts and bolts. As with the exterior ladder and the doors, these should be checked to make sure they are secure and in good repair.

8. Livestock

Livestock penned in feed rooms, feed lots, and around the silo area should be removed prior to fire operations. If animals become frightened, they may be impossible to control during fire operations, presenting a safety hazard to themselves and firefighters.

9. Hoops or Bands

The hoops, bands, or rods binding the staves are structural components and should not be used as ladders, footholds, or handholds. They are not designed for such use and have sharp edges, connections, and other potential sources of injury.

10. Structural Integrity

The main structural members of the typical silo will be exposed (such as the hoops and bands). These and other structural members such as the staves should be monitored for signs of weakness due to fire, wind, rain, ice, or other environmental condition that may cause structural failure. Remember that the weight of the silo and its contents may exceed 70 tons. Load-related structural failure of silos may cause significant injury and extensive property damage.

11. Exposure to Respiratory Toxins

In addition to the hazards caused by the oxides of nitrogen and oxygen depletion, a number of other silage-related respiratory hazards exist for emergency responders.

Pesticide exposure--Silage may contain the residue of various chemical treatments, which include pesticides and hay preservatives. Usually the pesticides are no longer present by the time silage is harvested and stored. However, the possibility of such exposure cannot be ruled out. If emergency personnel exhibit classic symptoms of pesticides poisoning (such as uncontrolled salivation and lachrymation, seizures, and other unexplained symptoms suggesting central nervous system damage), pesticide exposure may indicated.

Reaction to microorganisms---Tiny spores which grow in wet silage appear as white dust. This dust is actually a mold-like bacteria called thermophilic actinomycetes. Hypersensitive reaction to this bacteria is called "Farmer's Lung." Four to eight hours after exposure to the "dust," the individual exposed may develop flu-like symptoms (harsh cough, fever, chills, sweating, weakness, and shortness of breath). Farmers may build up a resistance to this bacteria. Those who are rarely exposed to this bacteria, such as emergency responders, may experience a sensitive reaction, resulting in an illness much like pneumonia. The term for this condition is hypersensitive pneumonitis.

Pulmonary mycotoxicosis---This condition is closely related to hypersensitivity pneumonitis, but is not associated with dust clouds of mold spores. This is caused by a fungus that develops in the top layer of exposed silage and causes a toxic reaction in the lungs. The symptoms are: burning in the eyes, throat, and chest; chills; fever; headache; and a dry, irritating cough.

Organic Toxic Dust Syndrome---Another respiratory illness, which may result from exposure to silage contaminated with microorganisms, is Organic Toxic Dust Syndrome (ODTS). This syndrome is characterized by fever and flu-like symptoms, shortness of breath, and impaired pulmonary function. An increase in white blood cell count may also be a sign of ODTS. It is commonly misdiagnosed as farmer's lung disease.

Two important cautions should be remembered about exposure to respiratory toxins:

- **Exposure to respiratory toxins can be prevented; always wear self-contained breathing apparatus in these atmospheres.**

- **In the event respiratory protection is not afforded, be very specific about information provided to medical personnel as to the symptoms and means of exposure.**

PRE-INCIDENT PLANNING

Pre-incident planning can provide critical information needed to safely and effectively combat silo fires. A fire department can and should identify the types and locations of silos in their response area and maintain such information at their station, communications center, and on the command vehicle.

The types of information needed are:

- contact information (telephone, pagers) for silo manufacturers, dealers, or others with specialized technical knowledge who will provide assistance during emergencies;

- location and types (conventional, oxygen-limiting, conversion);

- size of silo; dimensions and capacity;

- type(s) of silage generally stored;

- vehicular accessibility to silo;

- identification and protection of exposures; and

- water sources.

This information should be reviewed and updated periodically. It can be used for training and should be readily accessible response information.

FIRE SUPPRESSION OPERATIONS

As with any fire operation, firefighters should carefully evaluate the hazard and risk factors in silo fires. The fires are generally contained and may present no immediate danger. This is especially true of conventional silos. The proper risk assessment of the oxygen-limiting silo is: **Do not attempt to enter, breach, or otherwise attempt to extinguish fire in oxygen-limiting silos. There is nothing to gain and firefighters' lives are at considerable risk. Call the manufacturer's representative and protect exposures.**

Fires in conventional silos are slow burning, and, with the exception of embers falling down the unloading chute, are generally contained within the body of the silo. Fire departments can take the time not normally associated with more common types of fires to plan their response.

Firefighters should carefully consider the hazards, risks, and strategic objectives of silo fire operations. In determining the best tactical approach, these factors should be carefully considered:

- Life safety: always the primary factor. At most silo fires, the only lives in danger are those of the responders.

- Fire conditions: deep-seated, slow-burning, or rapidly developing, free-burning.

Exposures: distance, construction.

* Weather conditions: winds, humidity, electrical storms.

* Property: firefighters may prioritize the property to be protected in relation to proximity, value, and function.

* Protection of livestock: location and number.

* Resources: sufficient numbers of trained personnel and proper type and amount of equipment.

* Water supply: availability of sustained water supply, ability to deliver to the scene and apply to fire.

* Electrical services: transformers, wires, and disconnects.

The first step in planning the attack is to review the information acquired during the pre- incident planning phase. If this has not been done in advance, time must be taken at the incident to review the vital information and consider the age, type, and amount of silage loaded; weather conditions; time of day; and other incident-specific variables.

Some departments meet at the fire station to develop a plan of attack after acquiring the above information. This gives them an opportunity to review safety plans, incident command structure, equipment needs, staffing needs, and assign specific duties to responding fire fighters. (Some departments actually conduct a safety review prior to going to the scene for suppression operations.)

Size-Up of Incident

The extent and location of the fire can possibly be determined from the exterior of the silo as the following photograph indicates. The dry area around the unloading chute and below the loading door is an indication of fire. Outside observation can quickly indicate the dry area where the heat of the fire has dried the concrete staves. This pinpoints the location of the heaviest fire involvement. Size up may include observation of the interior via the outside ladder to the loading door platform.

Fire and smoke conditions must be carefully observed. Smoke may be issuing from the doors, chutes, or the dome area. The volume and force of the smoke as it exits the silo will be a good indication of the extent of fire within the silo.

Usually fire will occur in the top ten feet of silage, but it can and will burn deeper

Color differences indicate area dried due to heat of fire

around the unloading doors. To assist with ventilation of the silo, open as many doors as possible above the top of the silage.

For operations on and in the silo interior, this should be completed with the team entry and back-up procedure. Full turnout gear and SCBA should be standard personal protective equipment. Incident commanders may elect to make a limited visual survey of the interior to assess conditions.

Prior to initiating the entry and beginning the extinguishment phase, all command personnel, support staffing, and necessary equipment should be assembled. An essential decision process at this point is an assessment of hazard and risk. The incident commander must make a decision based on the stability of the structure, the extent of the fire, the availability of personnel and equipment, and the exposures involved.

In some cases, the fire may have left little or no salvageable foodstuffs and the condition of the silo may preclude entry for firefighting. A proper course of action may be to close all exterior openings and allow the fire to burn out. This can take a period of several weeks, and should be regularly monitored for flare-ups or exposure fires.

Staffing

One of the primary strategic and tactical considerations is staffing. Sufficient personnel should be available to support an operation that includes:

- **Command:**

  The Incident Commander should establish a command post at a safe vantage point to permit effective observation of the scene. The incident commander should remain visible to implement the command system and permit coordination of resources.

- **Safety/Observer:**

  The Incident Commander should designate a Safety Officer as early in the incident as possible. In silo fires, the Safety Officer should pay particular attention to the climbing and entry procedures, use of full protective clothing including SCBA, and the timing and accountability of firefighters working in the silo. The Safety Officer should also be sensitive to the requirements of rotating the firefighters in strenuous assignments.

- **Entry:**

  A sufficient number of firefighters to do the hands-on extinguishing and overhaul should be organized under an "Entry" or "Interior" sector officer. At least two firefighters should perform operations as a team inside the silo.

- **Back-Up:**

  A back-up or rapid intervention team of at least two fully equipped firefighters should be in close and constant visual or verbal contact with the entry crew. A good arrangement places an observer at the top of the silo who can observe and communicate with the entry team as well as alerting the back-up team to initiate support if needed.

- **Water Supply:**

  The Incident Commander should organize water supply operation to support a steady and uninterrupted flow of water for attack and overhaul. A sufficient reserve for emergency withdrawal coverage should be maintained as well.

All personnel should be accounted for at all times. Personnel must be rotated regularly if the decision is made to enter the silo, unload the contents, and extinguish the fire. This work is fatiguing and provisions should be made for crew relief and rehabilitation as well as for fluid replacement and food.

Water is the preferred extinguishing agent. Foam, whether it is protein-base, aqueous film forming foam, or flouroprotein is generally of little value in silage fires. The penetrating and cooling properties of plain water are superior to foam and most covering agents in silage fires.

Equipment

- Full turnout gear with self-contained breathing apparatus is required to operate in silos and chutes. Supplied air (airline respirator) can also be used by attack crews operating in the silo itself if such equipment is available.

- Firefighters in the silo should be equipped with safety harness and anchored to secure points via approved safety lines.

- A probing or piercing nozzle is useful. It must be of sufficient size and construction to penetrate four to six feet into silage for direct water application of at least 100 gallons per minute (gpm). Multi-directional nozzles such as a pierced-piping appliance can be used to great advantage to bore into the silage and apply water to the seat of the fire.

- Temperature probes, varying from two to six feet in length, can help pinpoint the location and extent of the burn area in silage layers.

- Lighting equipment should be used for exterior illumination and for interior operations.

- Ventilation fans may be employed for positive pressure ventilation.

- Sufficient radios and batteries will be required for extended operations.

- Auxiliary power will be required for light and ventilation.

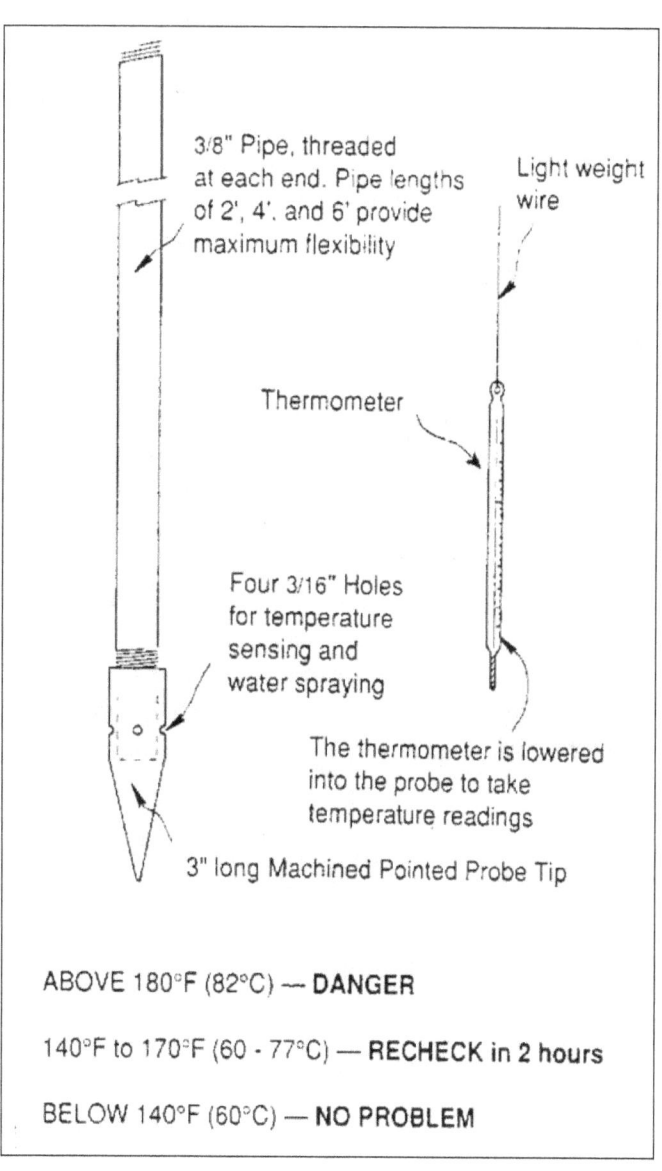

3/8" Pipe, threaded at each end. Pipe lengths of 2', 4', and 6' provide maximum flexibility

Light weight wire

Thermometer

Four 3/16" Holes for temperature sensing and water spraying

The thermometer is lowered into the probe to take temperature readings

3" long Machined Pointed Probe Tip

ABOVE 180°F (82°C) — **DANGER**

140°F to 170°F (60 - 77°C) — **RECHECK in 2 hours**

BELOW 140°F (60°C) — **NO PROBLEM**

Diagram 9. Temperature Probe

Tactics

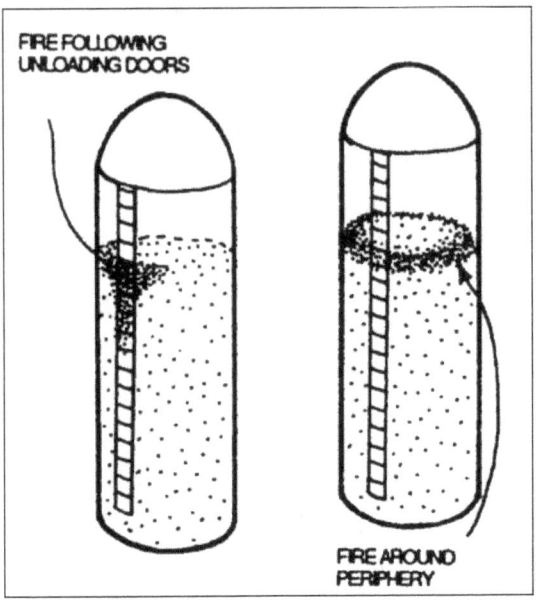

Diagram 10. The most common fire situation in conventional silos

When the decision is made to attack and extinguish the fire, the logical progression of knocking down visible exterior fire at the base should be the first priority. Positive pressure ventilation (PPV) should be applied at the base of the silo to clear heat, smoke, and gases from the fire area. PPV has been used to great advantage in these situations. However, firefighters must be alert for ashes or embers as well as increased fire intensity once PPV is utilized. Charged, supplied hose lines should be standing by before PPV is applied.

As crews enter the silo for extinguishment, great caution must be exercised. The burning silage may have created holes underneath that may not be visible. For this reason, entry crews should be tied off by a safety harness and line, and properly anchored. Ladders or other secure platforms can be used to distribute weight, but bridging of the contents can present very unstable situations.

The use of a temperature probe and penetrating nozzle may be used to good advantage. Probing should be done along both sides of the doors and around the periphery and near any other pockets exhibiting smoke or heat. The most common fire situations are noted in Diagram 10. These are typical fire patterns around the doors and the diameter of the silo. In some fires, both of these patterns will develop with burning around the exterior and fire also extending along both sides of the unloading doors in a "U"- or "V"-shaped pattern. Usually the hot spots in the silage can be easily identified through the resistance encountered by probes or by the reading from the temperature probes. Any area exceeding 180 degrees Fahrenheit should be injected with water.

Care must be taken if a flooding tactic is utilized. The bum pattern may extend from the ground up to the entire vertical length of the silo, creating an eccentric and unbalanced content load. As water is injected into the silage, it is absorbed and the dynamic weight of the load is greatly increased. In most silos, the content weight is designed to be transferred to the ground. Additional water loading increases the lateral load in an area that has little designed structural strength for lateral loading.

Diagram 11 depicts a firefighter using a probe to determine the location and extent of fire in the silo contents.

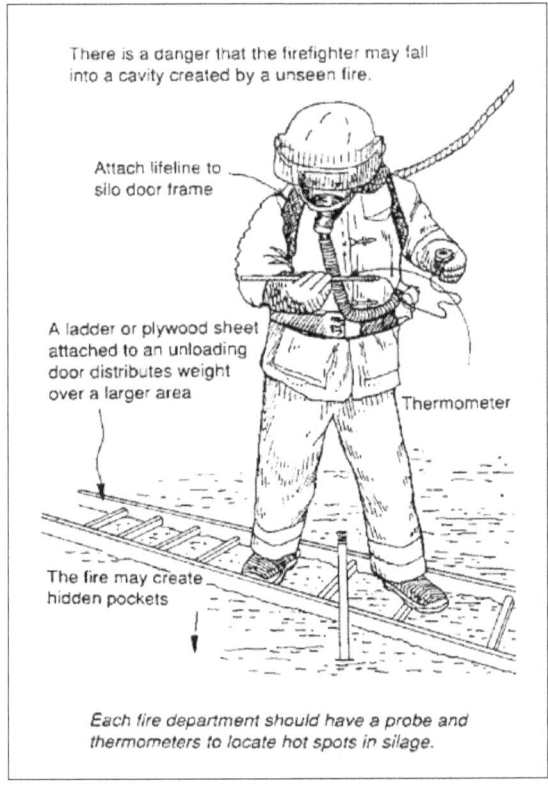

Diagram 11. Probing for hot spots in silage

OXYGEN-LIMITING SILOS

Under no circumstances should firefighters attempt to extinguish fires in oxygen-limiting silos. Rather, they should contact the manufacturer's representative or dealer. Extinguishment may be accomplished by injecting nitrogen or carbon dioxide into the silo using special fittings and piping available from the dealer. Firefighters must consider oxygen-limiting silo fires as a probable backdraft risk. Backdrafts in oxygen-limited silos have explosive force that can blast heavy structural components hundreds of feet away.

Cautions for fire in oxygen-limiting silos are:

- Do not add water or foam to the structure.

- Stay off these structures if smoke or steam is observed coming out of the roof openings (such as breather bag vents or pressure relief valves) and/or if rumbling and vibration of the structure is evident.

- Do not spray water **into** the structure or through any opening. Water will not penetrate or reach the burning material, but may draw air containing oxygen, which would cause a backdraft-like explosion.

- Do not spray water **onto** the structure as this will damage the structure itself and possibly cause an explosion. The water will not cool the structure, but will draw in air that could initiate an explosion.

- Assure that all unloaded doors, structure access doors, or any other openings are closed. **If the structure is rumbling and vibrating, no approach should be made. The fire should be allowed to burn itself out.**

BIBLIOGRAPHY

Thomas J. Maloney. "Combating Agricultural Silo Fires." Speaking of Fire, International Fire Service Training Association. Oklahoma State University. Stillwater, OK: Oct. 1997.

Silo Operators Manual. International Silo Association. P.O. Box 560, Lafayette, Indiana 47902.

Extinguishing Silo Fires. publication #NRAES-18. Northeast Regional Agriculture Engineering Service. Cooperative Extension. 152 Riley-Robb Hall, Ithaca, NY 14853-5701.

Ted Halpin, Doug Shattuck. "Agricultural Silo Fires." Fire Chief Magazine. H. Marvin Gin Corporation. Chicago, Illinois: September 1987.

"NIOSH Warns Farmers of Deadly Risk of Grain Suffocation." Publication 93-116. April 1993. National Institute of Occupational Safety and Health, Department of Health and Human Services. 200 Independence Avenue, SW, Washington DC, 20201.

"Preventing Fatalities Due to Fires and Explosions in Oxygen Limiting Silos." Publication #86-118. June 1986. National Institute of Occupational Safety and Health, Department of Health and Human Services. 200 Independence Avenue, SW, Washington DC, 20201.

Dennis J. Murphy, "Silo Gases-the Hidden Danger." Pennsylvania State University Fact Sheet #16. 1991. Pennsylvania Cooperative Extension Service, Pennsylvania State University, College of Agricultural Science, Agricultural Engineering Department. 246 Agricultural Engineering Building, University Park, PA, 16802.

Timothy G. Prather. "Silo Fires-Costly, Frustrating and Even Deadly." news release. August, 1993. University of Tennessee Agricultural Extension Service. Knoxville, Tennessee 37901.

Bradley K. Rein. "Farm Safety: Prevention, Rescue, and Rehabilitation." Document #0- 866-3 10. Farm Safety Fact Sheet. May 1991. United States Department of Agriculture Extension Service, Washington DC 20250-2260.

Charles V. Schwab, Laura Miller. "Electrocution Hazards on the Farm." Fact Sheet Pm-1265k. Safe Farm Program. November 1992. Iowa State University Extension, Ames, Iowa.

"Agricultural Dusts and Gases." Farm Occupational Safety Safety Information sheet. 1992. published by National Agricultural Library, United States Department of Agriculture. College Park, MD, 20850.

www.ingramcontent.com/pod-product-compliance
Lightning Source LLC
Chambersburg PA
CBHW081412170526
45166CB00010B/3311